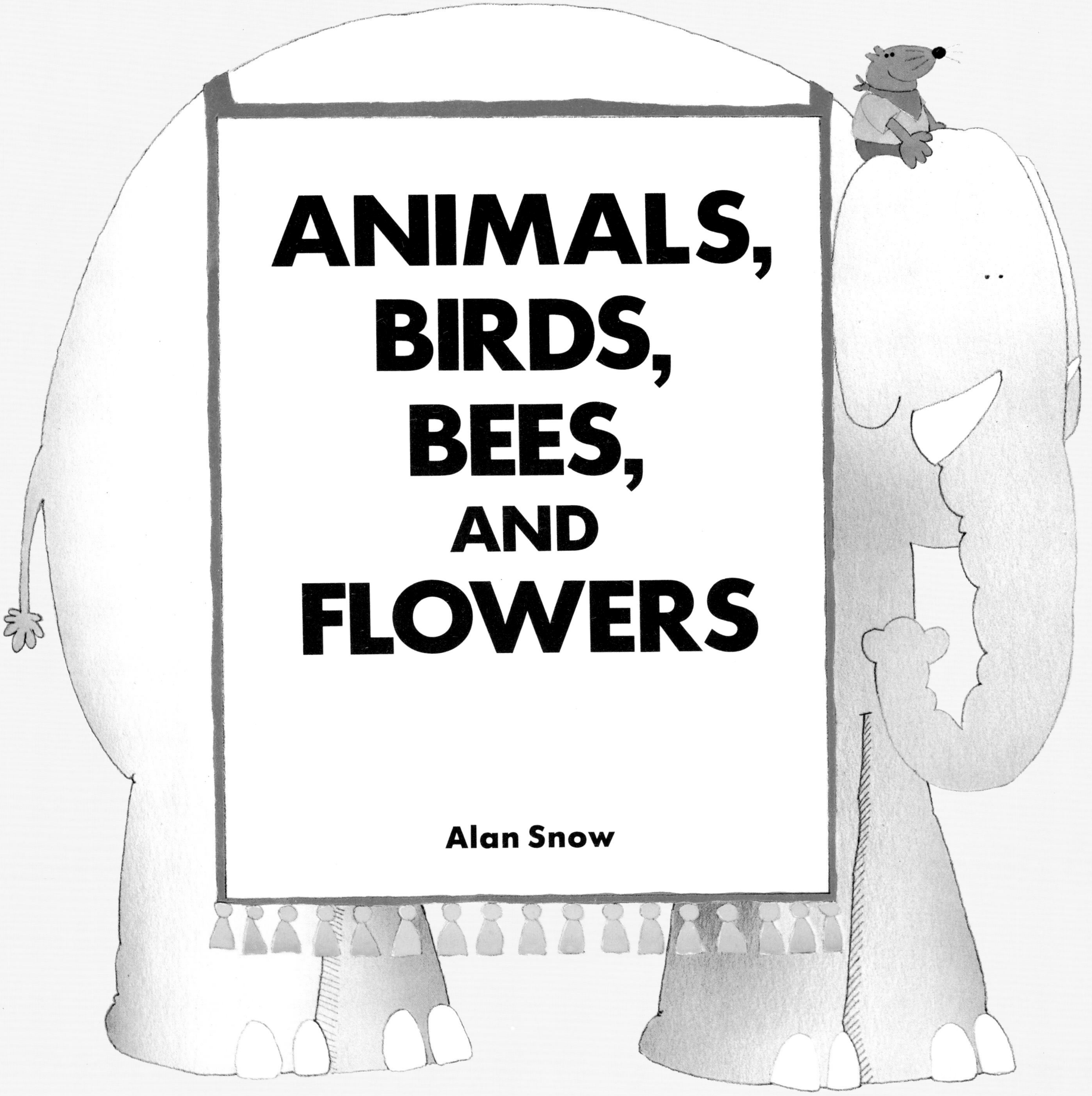

ANIMALS, BIRDS, BEES, AND FLOWERS

Alan Snow

Derrydale Books
New York

Dinosaurs

The first animals were reptiles. We call them dinosaurs, which means "terrible lizards." Some dinosaurs were huge, but others were very small.

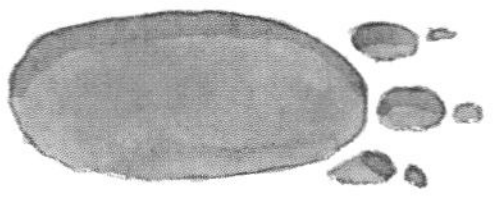

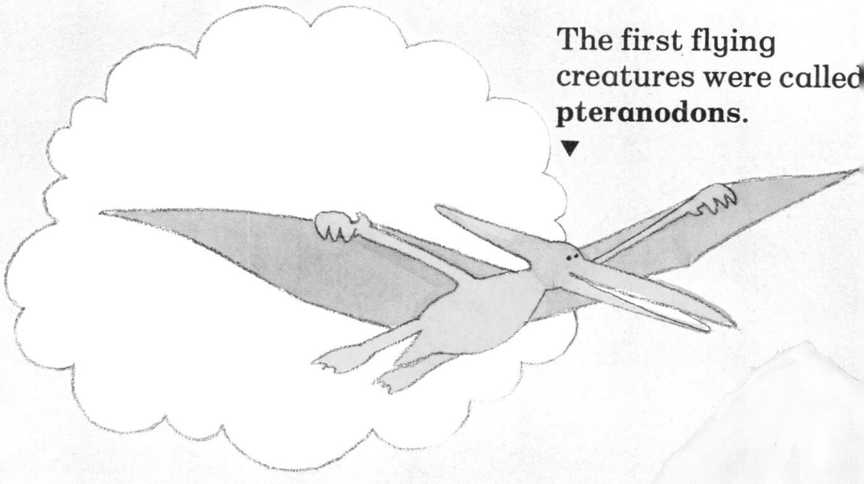

The first flying creatures were called **pteranodons**.

The first trees were called **giant horsetails**; some were as tall as an apartment building.
Munch, munch.
styracosaurus
triceratops
The **brontosaurus** was very big, but it ate only plants.
Crocodiles (**mesosaurus**) are a type of dinosaur. They look just like they did millions of years ago.
The first flying insects were giant dragonflies. They had wings more than a yard long.

Fossils are the remains of animals that have been turned into rock. Fossils help us to know what dinosaurs and other ancient creatures looked like.

Here are some fossils:

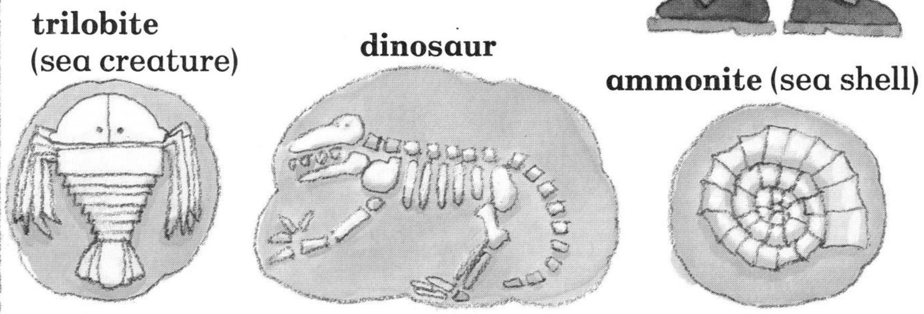

The **ankylosaurus** had thick, prickly skin to protect it from the meat-eating dinosaurs. ▼

▲ **Coelophysis** could escape from an enemy ▶ by running very fast.

Flowers

Flowering plants grow in nearly every part of the world. Plants need air, water, and sunshine to live and flower. Flowers produce seeds which grow into new plants.

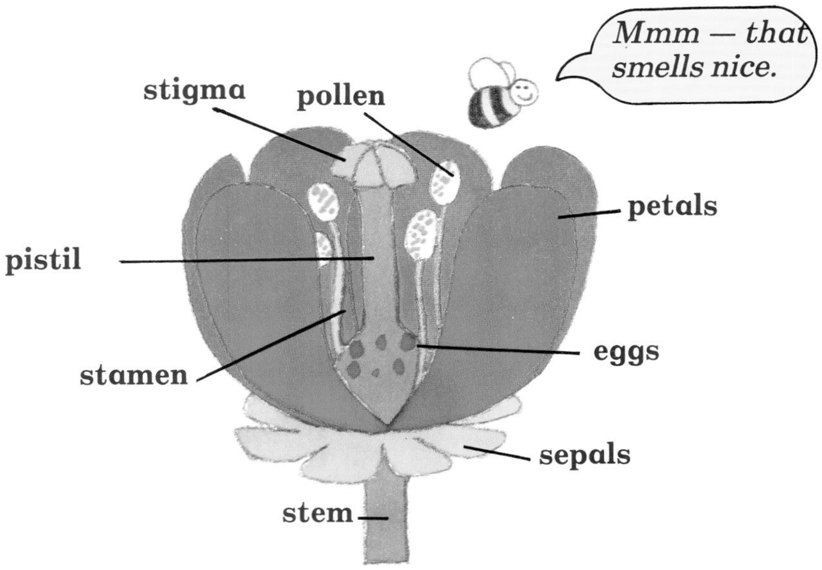

Here is a flower. Insects come to drink the nectar inside. They carry pollen from flower to flower. A flower needs pollen from another flower of the same kind before it can make seeds.

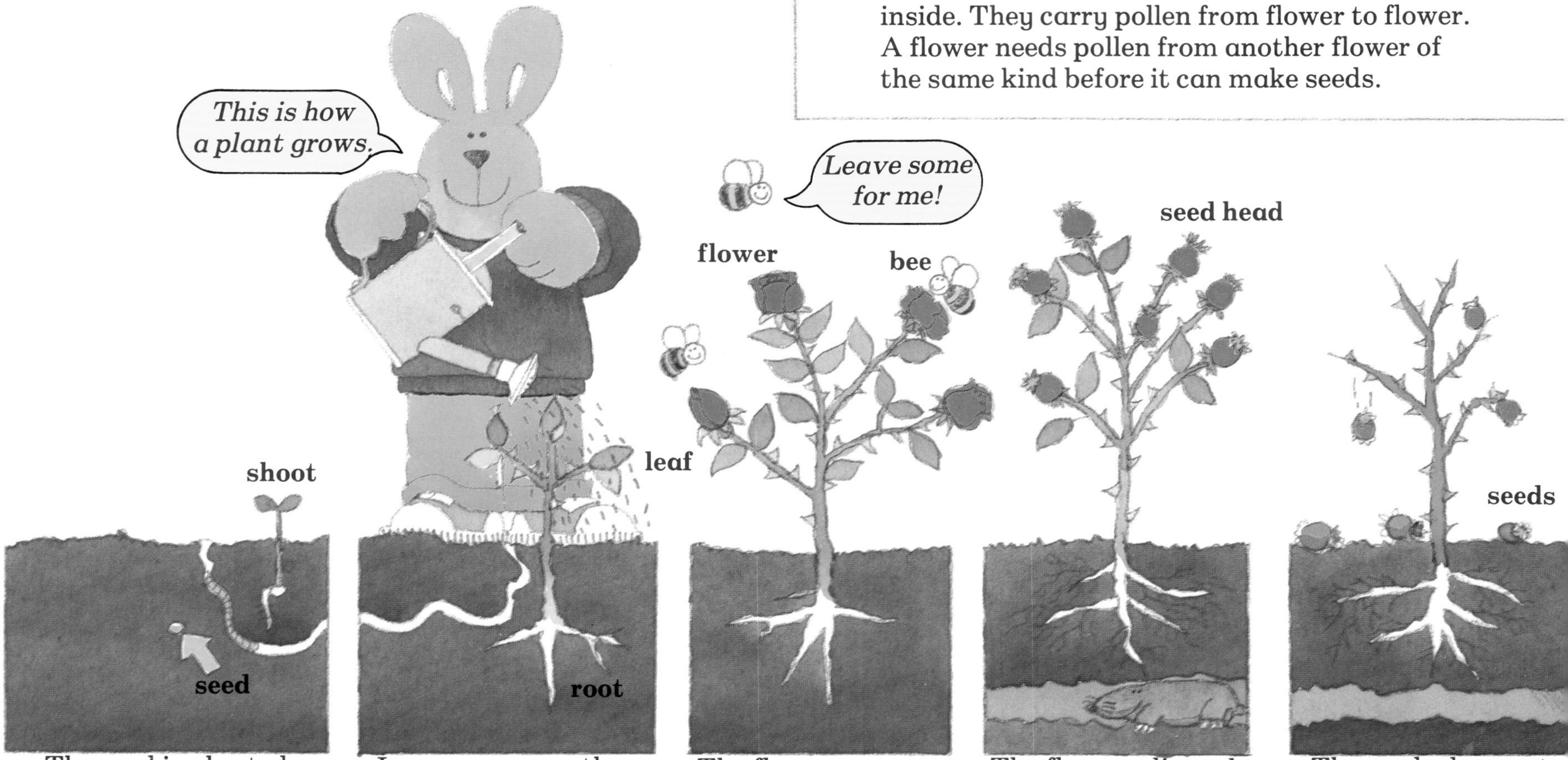

The seed is planted. It sends a root down and a shoot up.

Leaves grow on the shoot. The roots take in water.

The flowers open. Insects come to feed and pick up pollen.

The flowers die and leave seed heads full of seeds.

The seeds drop onto the ground. They grow into new plant

Strange plants.

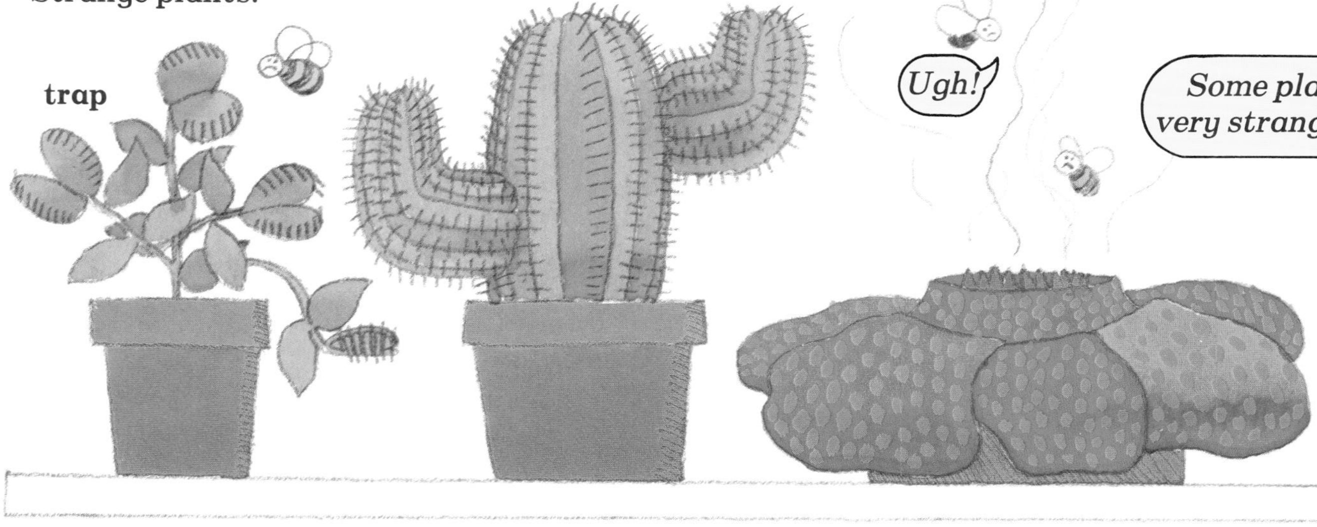

The leaves of the **Venus fly-trap** snap shut to catch flies for food.

The **cactus** grows in dry places, and keeps its own store of water inside its thick stem.

The **rafflesia** is the biggest flower in the world. It smells terrible!

Trees

Trees are large, woody, flowering plants. All over the world, trees are cut down for their wood. It is used to make things like paper, houses, and ships. New trees grow from the seeds formed inside fruits. These fruits can be apples, nuts, pods, or even pine cones.

The giant **sequoia** tree is the largest living thing in the world. ▶

Trees which keep their leaves in winter, like this **conifer**, are called evergreen.

Trees which lose their leaves in winter, like this **oak** tree, are called deciduous.

Trees make good homes. Many different animals, insects, and birds live in trees.

Insects

There are more insects in the world than any other kind of animal. Each insect has its own special skill. Some build nests and traps. Some can fly and others can jump. A few are poisonous. Insects have six legs, but some have more — spiders have eight and millipedes have dozens, while worms have no legs at all!

Bees live together in a huge family. The queen bee lays eggs. The worker bees make honey and build cells of wax in which the eggs and honey are stored.

Insects are very helpful to flowers. They carry pollen from one flower to another, and help them to make seeds.

Thousands of **ants** live in this nest. The nest is full of tunnels and spaces where eggs are kept. When the eggs hatch into baby ants called larvae, they are fed by the worker ants.

Worms burrow through the earth. As they burrow, they eat the soil for nourishment. Worms like to stay damp and cool.

Flies taste things with their feet! But their feet also pick up nasty germs and carry them around. Some flies even bite.

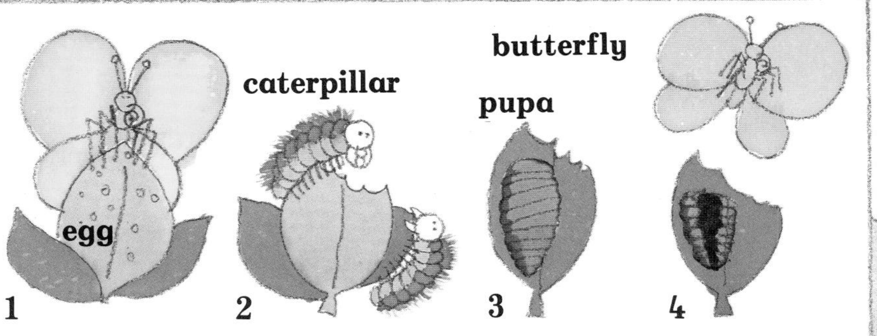

A butterfly grows up in four stages.

1. First it is an egg on a leaf.
2. The egg hatches and out comes a caterpillar.
3. The caterpillar changes into a pupa.
4. The pupa cracks open and the brand-new butterfly crawls out. After its wings harden it flies away.

Fish

Fish can only live in the water. They breathe it in through their mouths and out again through their gills. Fish move their tails from side to side to move through the water.

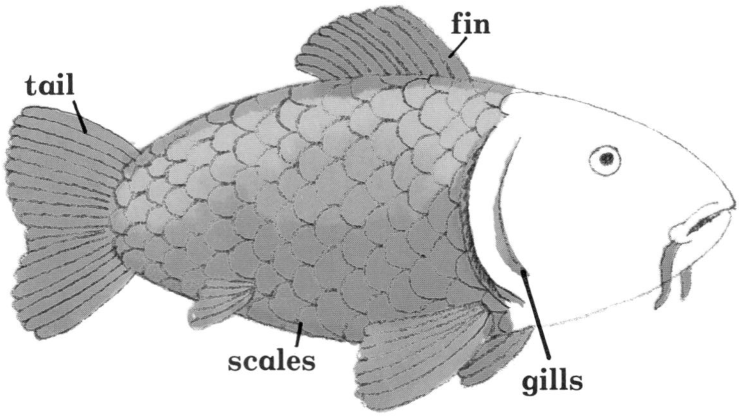

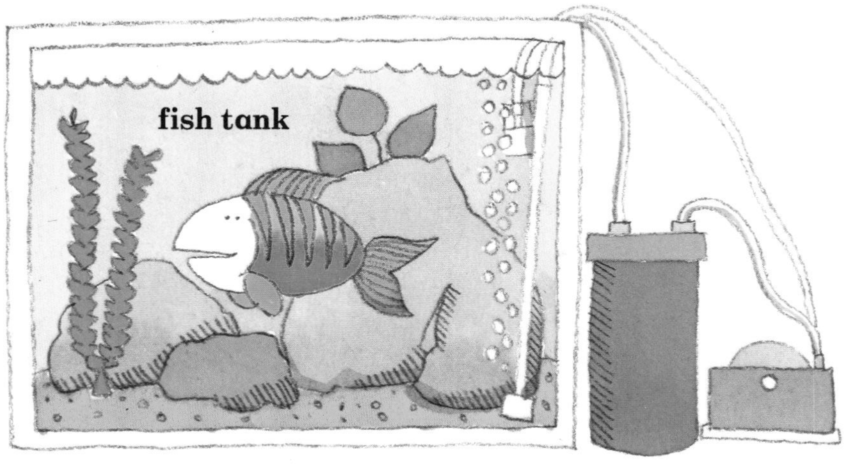

In the sea there is food for everyone. The small fish eat tiny animals and plants. The big fish eat the small fish. Humans and birds eat the big fish!

There are two main types of fish, those that live in fresh water, and those that live in salt water. Here are some salt water fish.

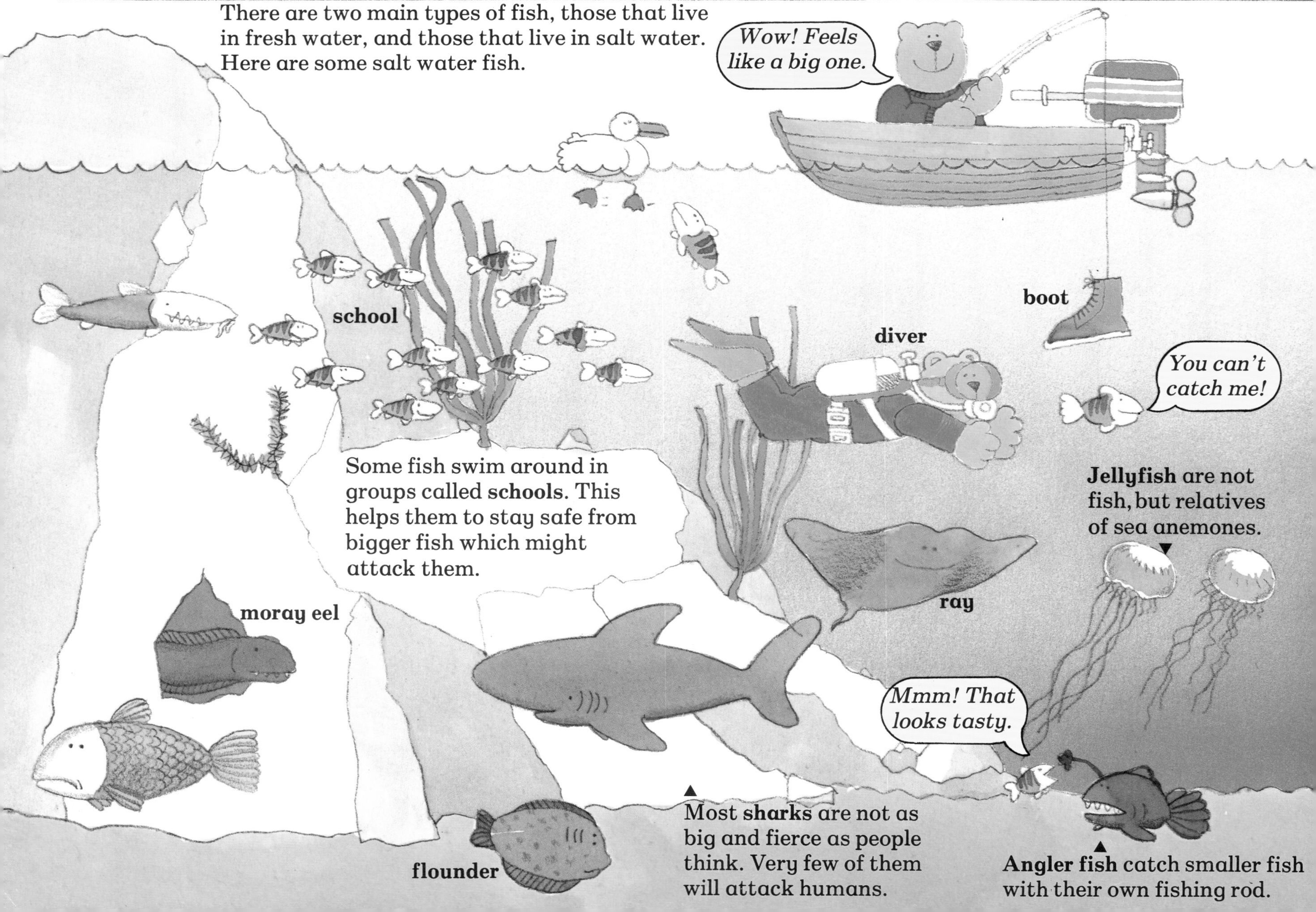

Some fish swim around in groups called **schools**. This helps them to stay safe from bigger fish which might attack them.

Jellyfish are not fish, but relatives of sea anemones. ▼

▲ Most **sharks** are not as big and fierce as people think. Very few of them will attack humans.

▲ **Angler fish** catch smaller fish with their own fishing rod.

Reptiles

Reptiles have a tough, scaly skin. They lay eggs on land, which hatch out in the warm sun. Turtles, snakes, and lizards are all kinds of reptiles.

Amphibians live in the water when they are young and on land when they grow up. Frogs, toads, and newts are all amphibians.

A frog grows up in four stages.

1. The female frog lays eggs in the water which are quickly fertilized.
2. The eggs hatch and tiny tadpoles wiggle out.
3. The tadpoles grow legs and change into tiny frogs.
4. The frogs then crawl out of the water.

Birds

Birds are covered in feathers, which keep them warm and dry. The feathers are very light, and they help the birds to fly. Because birds can fly, they can live almost anywhere.

The **condor** is the largest flying bird. ▼

Bird watching is a favorite hobby. People who watch birds often build shelters, called blinds, where the birds can't see them. ◄

blind

cliff

Many seabirds nest on high cliffs. No one can reach them there, so they are safe. ▲

beach

Swifts spend most of their time flying. They swoop about catching insects in their mouths. ► Some swifts may stay in the air for a year!

cormorant

Bird watching is fun. Why don't you try it?

puffin

▲ **Penguins** cannot fly, but they can swim underwater and catch fish.

seagull

sea

Not all birds' **feet** are the same.

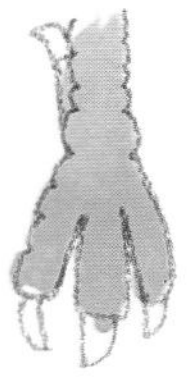

An eagle has strong talons for gripping its prey.

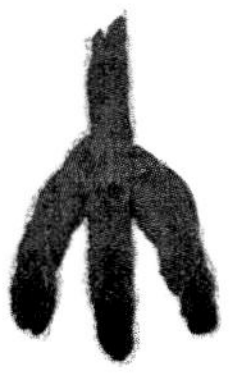

A magpie has toes to help it walk or hop on the ground.

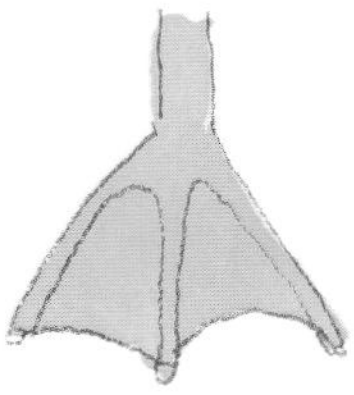

A duck has webbed feet so that it can paddle in the water.

Birds' beaks come in many shapes and sizes. This helps them in gathering and eating their food.

A sparrow has a short, stout beak for picking up tiny seeds from the ground.

A parrot has a big, strong beak for cracking large nuts.

An eagle has a curved beak for tearing flesh.

Many birds make their nests in trees. They often build them with twigs and grass. Then they may line the nests with cotton or moss to make them cozy. ▼

How a bird grows.

The mother bird lays eggs in the nest.

The chicks hatch.

The parents bring food to the nest.

The young birds learn to fly.

Mammals

Mammals come in all shapes and sizes. They have fur or hair to keep them warm. All baby mammals drink milk from their mothers as their first food. Most mammals live on the land, but bats can fly and some squirrels can glide. Whales and dolphins are mammals, even though they live in the sea.

Here are lots of different mammals. How many can you name?

1. camel
2. elephant
3. chimpanzee
4. bush baby
5. giraffe
6. bat
7. horse
8. cow
9. kangaroo
10. goat
11. oryx
12. donkey
13. sloth
14. hippopotamus
15. rhinoceros
16. buffalo
17. whale
18. moose
19. badger
20. capybara
21. pig
22. dog
23. sheep
24. fox
25. wolf
26. tiger
27. lion
28. grizzly bear
29. polar bear
30. panda
31. rabbit
32. cat
33. koala
34. boar
35. leopard
36. mole
37. otter
38. beaver
39. squirrel
40. mouse
41. seal

Farm

Farmers look after animals and feed them. In return, animals give us many things, such as wool, eggs, milk, and meat.

Sheep have thick, wooly coats called fleeces. Once a year the farmer cuts off the fleeces. They are taken to a factory and spun into wool. The wool is used to make clothes for us to wear.

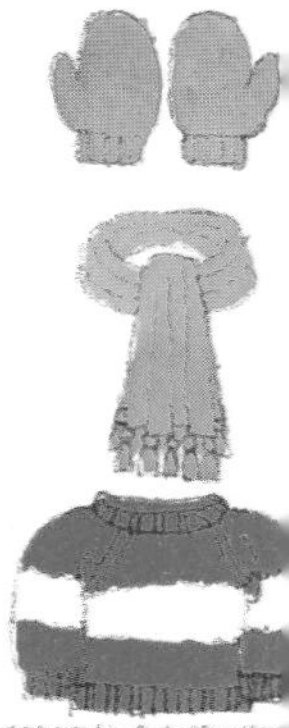

ram
Baa!
sheep
A baby sheep is called a **lamb**.
crook
ewe
shepherd
The **sheepdog** helps the farmer to round up the sheep.
wire
fence post
chicken house
Cheep!
feeder
A baby chicken is called a **chick**.
Cluck!
egg
chicken
A chicken lays an egg nearly every day.

Farm animals were once wild animals. This is what they looked like long ago. More and more unusual animals are now being kept as farm animals, such as deer, kangaroos, and even fish!

Cattle had long horns.

Pigs had tusks and hard, bristly coats.

Sheep had long coats.

Pigs often live in open fields. They dig in the earth with their strong noses. In hot weather they bathe in mud to keep cool. But most of the time they are very clean animals.

Dairy cows are kept for the milk they give. The cows must be milked twice every day. You can drink milk or eat foods which are made of milk. Here are some of them:

On Safari

Many large animals eat grass. They live on big, open grasslands, eating all through the day. But meat-eating animals live there too. They hunt the grass-eaters for their food.

Rhinoceroses look b
and slow. But when
they are angry they
become fierce and c
run very fast.

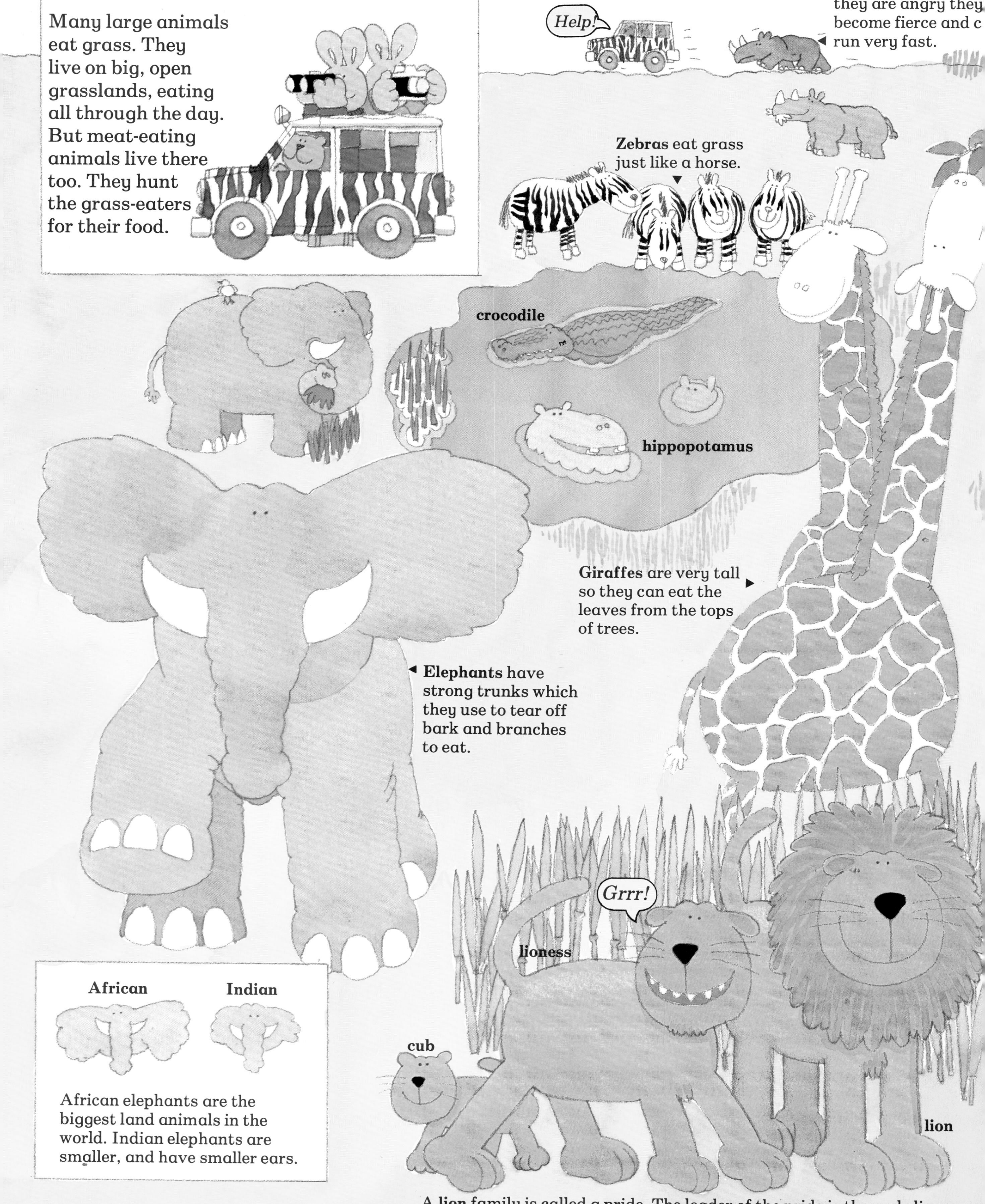

Zebras eat grass just like a horse.

Giraffes are very tall so they can eat the leaves from the tops of trees.

Elephants have strong trunks which they use to tear off bark and branches to eat.

African elephants are the biggest land animals in the world. Indian elephants are smaller, and have smaller ears.

A **lion** family is called a pride. The leader of the pride is the male lion. But the lionesses do most of the hunting.

In the Jungle

Jungles are hot, wet forests where trees and other plants grow very quickly. Jungles are the home of many kinds of animals.

Eagles fly high above the jungle trees. They swoop down to catch their food in the tree tops.

Leopards often climb and hide in trees. They ambush deer by jumping down on them. After their meal, they climb up again — and go to sleep!

Monkeys live up in the trees, where they make a lot of noise! With their long arms they can swing through the branches at high speed.

In the Ocean

Most of the world is covered by oceans. Fish live in the ocean as well as many strange creatures. There are whales, squid, crabs, and dolphins. Other animals only hunt for food in the ocean while they live on land.

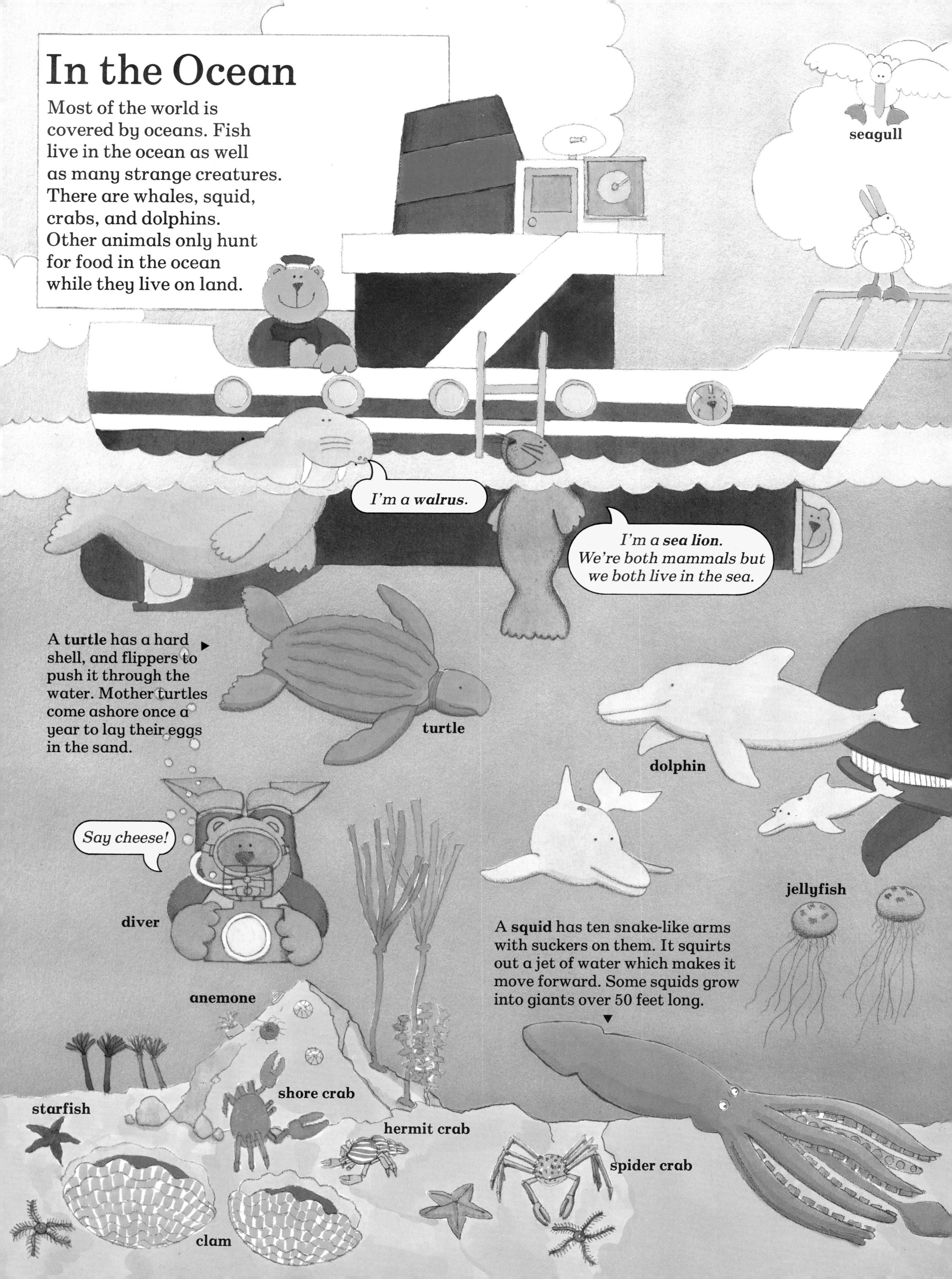

A **turtle** has a hard shell, and flippers to push it through the water. Mother turtles come ashore once a year to lay their eggs in the sand.

A **squid** has ten snake-like arms with suckers on them. It squirts out a jet of water which makes it move forward. Some squids grow into giants over 50 feet long.

Coral reefs are made from the hard shells of tiny coral animals. Many beautiful fish live in the reefs.

Polar bears are land animals but they can swim very well. They hunt for fish and seals under the water.

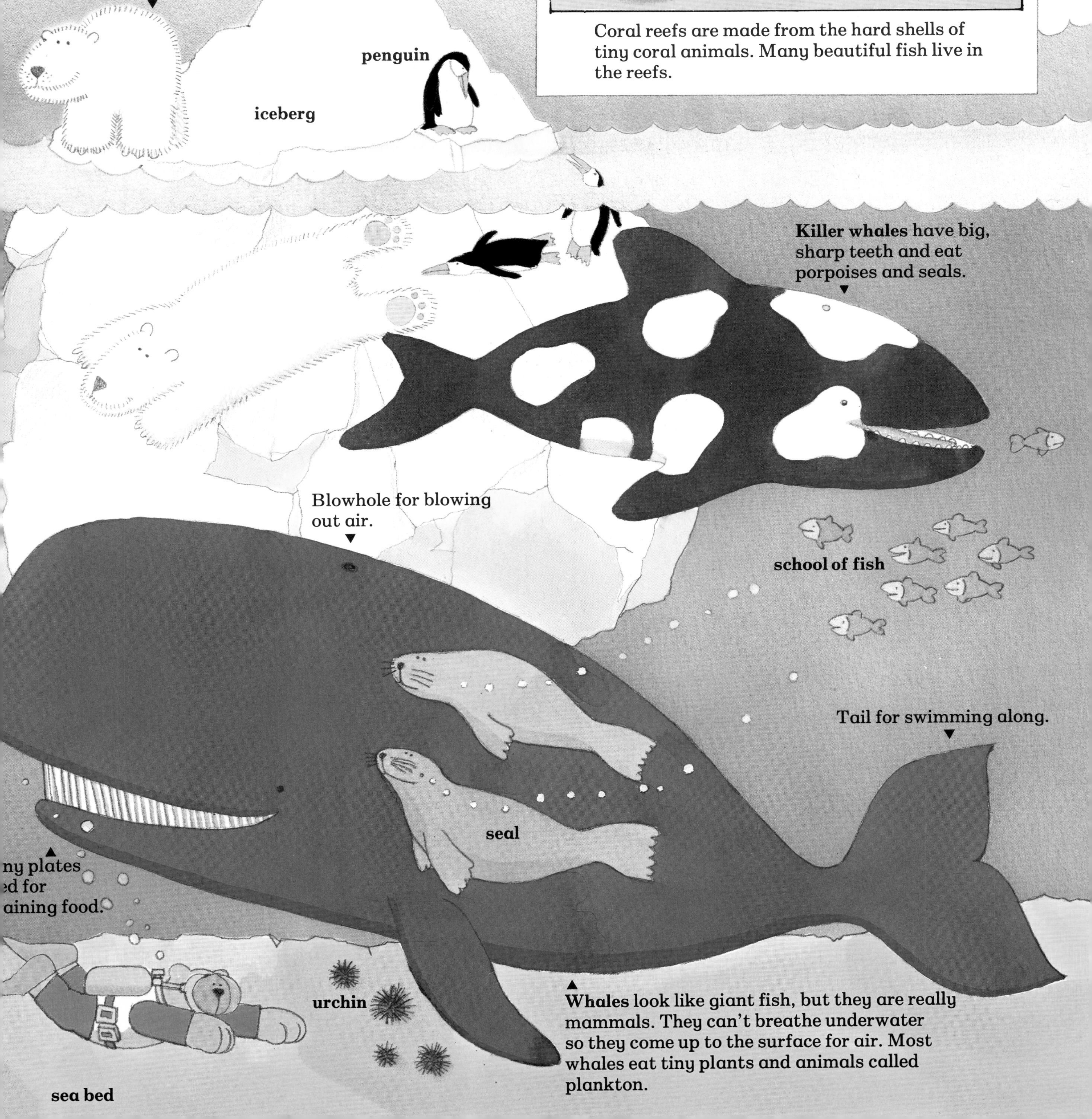

Killer whales have big, sharp teeth and eat porpoises and seals.

Blowhole for blowing out air.

Tail for swimming along.

ny plates
ɜd for
aining food.

Whales look like giant fish, but they are really mammals. They can't breathe underwater so they come up to the surface for air. Most whales eat tiny plants and animals called plankton.

At Home

Wherever you live, there are animals and plants living nearby. See how many you can spot in the woods, fields, and river on this page. How many can you name?